THE CITY OF THE NUMBER GAMES

An Amazonian Adventure

Cacildo Marques

ISBN: **978-1547226443**

Marques, Cacildo
The city of the number games: An Amazonian Adventure/ Cacildo
Marques.
EpistemeEd , 2017.

54p.
ISBN: 978-1547226443

I. Arithmetic Operations - Games. I. Titles   CDD 513.2

CONTENTS

Little warning

When his first daughter was born, Wiliam Gates III, Bill Gates, being asked about when he would give her a computer, he replied: When she knows enough English and Mathematics. Those who buy electronic technology products for their children should follow the example of the Microsoft founder. And for learning Math softly, one way is to practice number games, far away from the electronic machines.

Let the computer games for those who already know enough English and Math. For the youngest, we will give pencils, notebooks, handouts and books, preferably fun books.

Small children with a tablet in their hands are holding up a lot their domination of speech. Decades ago, they started talking in one year old, but now this occurs at 30 months, foreshadowing other problems. Let's get the electronics out of their way, as some conscientious psychiatrists are recommending. The healthy one is to give crayons, scribble paper, modeling dough, elastic carts, dolls and everything from which children have always liked.

My number game book did not get any idea of other games. They are all unpublished so far. Many hundreds of numeric games exist scattered around the world, and mine come to add new challenges. They are appearing inside a story, which is happening in a city in the Amazon.

Enjoy!

The author.

THE CITY OF THE NUMBER GAMES
An Amazonian Adventure

Cacildo Marques

1. THE CITY OF THE NUMBER GAMES

The rain had already passed when the small airplane taking Teacher Lalo Vieira and his son Marino landed in the city of Guarana, headquarters of the municipal district-reservation of a Tupi branch tribe in the State of Amazon.

In other times Guarana was just a village inhabited for a reduced number of natives who dominated some words of the Portuguese language and one or another of English, words learned with foreign travelers who passed a certain time there. One of those travelers, Teacher Edmond Smith, left a deep mark among those people when, some decades ago, he left his home town, inside Great-Britain, where he acted in the medium teaching, and came to Guarana with the objective of doing field studies for his thesis of doctorate in linguistics. But, differently of others who just sought to transmit religious doctrines or to compile the natives' knowledge, mainly on the forest, Teacher Edmond also had the purpose of winning followers for another area of the knowledge: the Mathematical Science.

Lalo Vieira knew the history of Teacher Edmond's passage for Guarana. He knew that he had left in the middle of that people the healthy habit of playing with numbers. Starting from that stay, the small children desired to learn the first arithmetic operations not just to prepare for the practical life and for the subsequent studies, but also to participate in the games that represented the more disseminated amusement among the majority of the children.

The objective of Lalo was to write a work for the university,

1

describing the way how the little indians of Guarana lived together with the habit of playing with number games, with the tradition of hearing histories and the amusement of the practice of sports. Marino, who was 13 years-old, was taken together because certain aspects of a people culture can only be captured by a child or an adolescent.

The mayor, Mr. Tieh, went to receive the visitors in the airport. Close to him, his children Amana and Iandu walked, and they soon made friendship with Marino and showed him, already in the itinerary toward the mayor's house, an object that woke up the curiosity not only of the son, but also of the father, the teacher. It dealt with the "big die", an numbered icosahedron that is used to do raffles and many types of games. The icosahedron is a solid of 20 faces and, when it is regular, those faces are equilateral triangles. Faces being triangles, the figure is one of the five polyhedrons of Plato. The tetrahedron, the cube, the octahedron and the dodecahedron are the other four ones. In the big die, the faces are numbered from 0 to 9 twice, being once in blue and other in red. In the games that involve positive and negative numbers, the red cipher has negative value and the blue, positive. When there is no need to distinguish between positive and negative, then one ignores the color of the ciphers, being considered all as no-negative. As one can notice, it is a die that supplies, in each throw, a cipher between 0 and 9, with equal probability. With this, the values of the numbering system are had in base ten, what does not happen with the traditional die.

Returning to the house, Iandu showed Marino the process of construction of the big die. In general it is built in cardboard

2

and the one that one needs to know is the planning of that solid. One deals with a drawing with 20 chained equilateral triangles, 10 of them being in the array of the middle, 5 in the superior part and 5 in the inferior part.

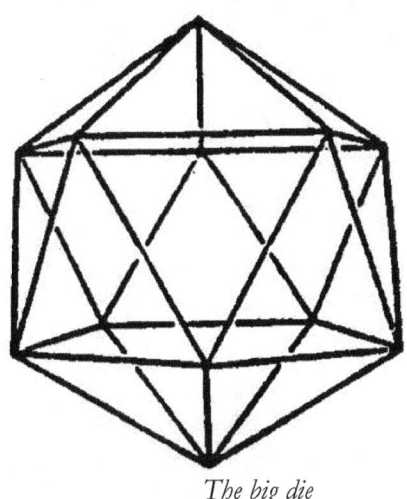

The big die

Any skilled child, with some education, can build the big die just using cardboard, pencil, ruler, compass, blue and red pens and glue.

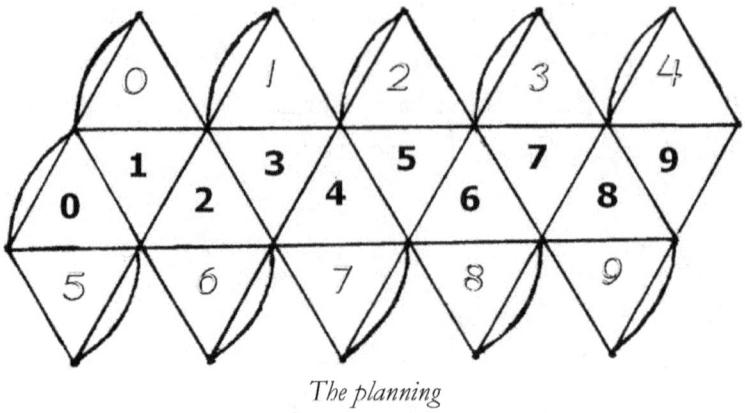

The planning

First a straight line is drawn and, with a convenient opening of the compass, which will have the size of the triangle base in each face, chained circles are drawn, one following the other. Then, semi-circles are drawn in a second floor and, finally, the triangles are drawn and so the planning of the big die is had. Before cutting out and gluing it, one should write the numbering of the faces.

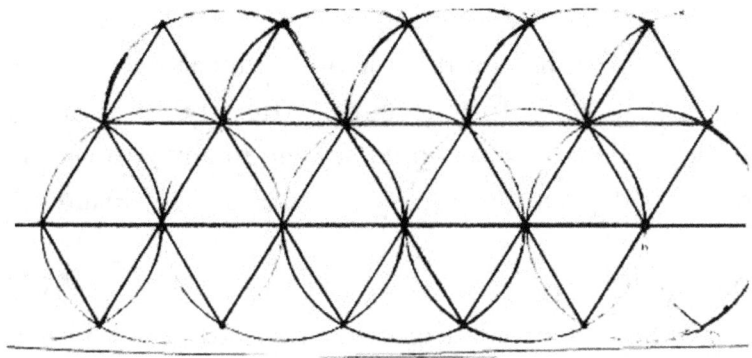

The sketch of the planning

Marino received those instructions from Amana and Iandu and in a few time he was already designing the drawing and setting up the solid that the Guaranaensis calls big die. As Iandu had explained, the size of the big die was by criterion of whom will build it and Marino built first one that possessed for each face a base with approximately three centimeters.

After agglutinating the tips cut out in the triangle with aim to form the regular icosahedron, Marino left his work to rest until the glue was very dry and very glued. Meanwhile it was arrived the lunchtime and the mayor's wife came to call her two children and Marino for the table. After they finished eating lunch, there, then, Marino would ask Iandu and Amana to teach him some of the number games so that he could inaugurate his big die.

During the lunch time, mayor Tieh invited Lalo to visit the local school. Marino would go together and he could pass the afternoon attending the classes with Amana and Iandu. Of course this was in Teacher Lalo's plans, once he intended to know the process of teaching of the city.

While Iandu got ready at his room, he went teaching to Marino the first number game, inaugurating the big die that was just built. Iandu explained that the game is very simple and it just involves the operations of addition and multiplication of whole numbers, and one more closure of parentheses, because it is the "game of the numerical expression".

Iandu also explained to Marino that, when there are no negative numbers evolved in the game, the big die can be substituted by ten tiles of the domino: 0x0, 1x0, 1x1, 2x1, 2x2, 3x2, 3x3, 4x3, 4x4 e 5x4. They represent the numbers 0, 1, 2, 3, 4, 5, 6, 7, 8 and 9.

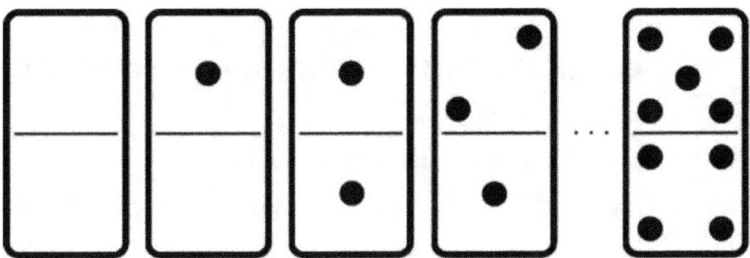
Tiles 0, 1, 2, 3, …, 9

2. THE GAME OF THE NUMERICAL EXPRESSION

That first game is made as it follows below.

Step 1: Each participant writes down in a paper the outline below:

$$...+ ...\bullet (...+ ...\bullet...)$$

The stippled spaces will be filled out with the ciphers that will go appearing in the superior face of the big die in its successive throws.

Step 2: The participants of the game go throwing the big die, one in every time, until each one obtains the 5 numbers and fill out his expression. The numbers in red or in blue are here considered non-negative, indistinctly.

Step 3: Each participant solves his expression and passes it later to be checked by the other player.

The player who obtains a larger value as a result of the expression wins the set, which can be repeated indefinitely.

Marino knowed it, but Iandu asked attention for the rules: in the expression, first one solves the parentheses, and, among the operations, one solves firstly the multiplication, and, following this, the addition.

Marino then told Iandu that he would take his big die to school and that in the break the two boys would play some sets of the numerical expression game. Iandu agreed in full with the idea.

Arriving in the school, Teacher Lalo and the mayor was

talking with the principal while Marino drove to the classroom of Iandu, together with this one. Amana, who was not of the same grade, went to another room.

The three classes before the recreation passed very quickly. They were classes of geography, music and mathematics and Marino so much wrapped up that already didn't remind that he had arrived at the school anxious for the break to play using his big die. The more appreciated class was the one of mathematics, favorite subject matter of Marino, but he had also liked the music class. As for the geography class, he admitted to Iandu later that he preferred the teacher, who was a nice girl. In the recreation, they joined with other two friends and accomplished several sets of the numerical expression game. Returning later to the classroom, they had the history class, which Marino appreciated a lot, and the ones of sciences and Portuguese language.

The time elapsed very fast on that afternoon, as a sign that Marino had liked really the classes. But it would not be the case of attending the classes of all the following days, because no longer they would be an innovation, it is what he imagined.

The one that Marino wanted was to learn new number games and this happened soon at the following day when Amana, early in the morning, came to teach him the game of the divisibility.

3. THE GAME OF THE DIVISIBILITY

The game develops according to the way below.

Step 1: The two players choose even or odd, a choose whose result will be maintained until the end of the game. Soon afterwards, each chooses a number of 5 ciphers, which can be obtained in the big die, and the two players sum those two numbers.

Step 2: If the sum is an even number and the result of casting out nines of that number gives 0, 3 or 6, then, the player who chose even number gains the points. If the result is odd, the points go to the player who has chosen odd, provided that the "out nines" is 0, 3 or 6. (the result of casting out nines for a given number is the sum of its digits up to the obtaining of a value smaller than nine. If the first sum is larger than nine, it is made out-nines of the out-nines. The out-nines of the number 9 is obviously 0.)

If the out-nines is different from 0, 3 or 6, the player who lost in the even-odd gains 1 point (or he adds 1 point to what he already has).

The points of the winner are counted like this: One takes the sum of the digits of the sum of the given numbers. One divides that sum by 3 or by 9 (by 9 if the out-nines gives 0 and by 3 in the other two cases, i. e., 3 or 6). The quotient obtained so is the number of points obtained in the throwing.

Step 3: The players write, each, a number of 4 ciphers and add the two numbers. With that new result they repeat the process used in the case of the 5 digits.

Soon afterwards they make the same thing for numbers of 3 ciphers, after 2, and, finally, of 1. In this moment, he who has more points wins the game.

It follows an example below.

Player A chooses even and player B chooses odd.

		Sum of the ciphers	Out-nines	
1st throwing	A:	41274		
	B:	10000		
Total:		41274.........	19...........	1
2nd throwing	A:	3415		
	B:	0212		
Total:		3617..........	18...........	0
3rd throwing	A:	219		
	B:	101		
Total:		320............	5..............	5
4th throwing	A:	..12		
	B:	..98		
Total:		110............	2.............	2
5th throwing	A:	..5		
	B:	..7		
Total:		12..............	3.............	3

Result of the 1st throwing: The number (Total) is non-divisible by 3 or by 9. The player B won in the "even-odd", but he lost in the divisibility. The player A, then, is with 1 point.

Result of the 2nd throwing: The number is divisible by 9 (because it gave 18 in the sum of ciphers). The player A won in the divisibility. He divides then 18 by 9 and obtains 2 points.

Result of the 3rd throwing: The number is non-divisible by 3 or by 9. The player B won in the "even-odd", but he lost in the divisibility. The player A increases 1 point to his.

Result of the 4th throwing: The number is non-divisible by 3 or by 9. The player A won in the "even-odd", but he lost in the divisibility. The player B gains 1 point.

Result of the 5th throwing: The number is divisible by 3 and by 6 (it has sum 3 and it is even) and the sum is divisible for 3. The player B won in the "even-odd" (the sum of the ciphers is odd) and also in the divisibility. The player B gains 1 new point.

Final result:	1st	2nd	3rd	4th	5th	Balance
A:	1	2	1	0	0	4
B:	0	0	0	1	1	2

Another modality of the game of the divisibility consists of doing the division not with the sum of ciphers of the Total, but with the own Total, dividing it by the out-nines. In this case, the divisions will take more time. Here, if the player loses for the number to be non-divisible, the other gains not 1 point, but 1000, 100, 10, 1 or 0, as he is in the 1st, 2nd, 3rd, 4th or 5th throwing. In one and in another modality the division is not used by 6, because this would alter the probability against the player who chose even.

Marino, after playing with Amana and with Iandu, who has appeared soon then, went to look for his father, Lalo, to play.

They made several sets and the teacher lost the most of them for Marino. "It is a game of pure luck" – Lalo argued, - "and you, son, have more luck than I".

4. A FEMININE PRESENTATION

Marino and Teacher Lalo was invited to go to the school on that Saturday.

- But, how is it? Do you have classes on Saturdays? - the teacher asked. There were not classes on Saturdays. The mayor explained that in the school the huka-huka, a fight among women, would be happening. It was like a kind of ritual, that the school students relived once per semester, to maintain the tradition of the tribe. The organization of the event was under the responsibility of the physical education teacher, who provide to the city many moments of amusement during the academic year.

Physical Education, Mathematics and Music formed the group of subject matters more appreciated by the pupils of Guarana. Teacher Lalo commented about this that those were three matters with a common characteristic: the necessity of training and orientation, besides to be the matters founder of the gym idea. That was very visible in the moment of the students' presentation in the huka-huka fight, because they demonstrated to be in shape, exhibiting a corporal agility out of the common. The fighters dressed a singular type of bikini: the bra was just a cloth ribbon tied in turn of the thorax. It happens that the bra usually used by the Guaranaensis youths was of the type cup, just containing the inferior part, to leave free the part from the breasts. That is what the mayor explained to Lalo, adding that the reason was the full development of the body. Using the bra without the top part, the girls of the city impeded that the breasts became flaccid, for holding them, at the same time in that they allowed their full development, doing not atrophy them. It is so that for the bikini of the huka-huka it was necessary to use a special piece, covering the breast entirely, since the usual bra would not be adapted for the occasion.

In the last round of the huka-huka, Marino again got to beat his father. The fighter for whom he decided to bet won the fight, defeating the one who had been the favorite of Lalo. Once again Lalo went out with the argument that Marino had more luck than he did. Tsk, tsk, Marino did: the one that happened is that his intuition had been sharper than the one of the father. Lalo gave a laughter and ended half lonely: "Indeed, maybe you are right".

At night, Marino would have one more game to learn from Iandu and he would play again with Lalo, after practicing it with his Guaranaensis friend.

5. THE GAME OF THE BASE TEN

This game is made with the domino and aims to help the practitioner to fix the concept of positional system in base ten.

Initially, an expression is written in the polynomial form of the numbers and then one looks for filling, one to one, its coefficients, which must be obtained in successive withdrawals from the domino. The expression is: $1 * 10^3 + ... * 10^2 + ... * 10^1 + ...$

The first coefficient, of higher power, has already been determined, and has value 1. The other three are being drawn.

To facilitate the accounts, each participant writes his expression already with the powers calculated:

$$1*1000+...*100+...*10+...$$

Step 1: After deciding who makes the first withdrawal in the domino, the participant who will not do it first, shuffles the bones. The other, player A, removes his tile, placing it in longitudinal position in front of himself, turns it and reads the value of the base. For example, if the tile is 3x2 and the 2 is facing the player, his value is 2. He writes this value in the house of the unit. If the value obtained is white, or zero, the player has the right to make one more withdrawal, replacing the previous value. If this second withdrawal returns to zero, that value is valid.

Step 2: Player B makes his withdrawal and also writes the

value in the unit house of his expression.

Step 3: Player A makes his withdrawal for the coefficient of the tens. He writes it in the position indicated in his numerical expression.

Step 4: Player B also obtains in the domino his coefficient for the position of tens and writes the value.

Step 5: Player A makes his last withdrawal, filling in the coefficient of the house of hundreds.

Step 6: Player B also makes his final withdrawal and completes his expression.

Step 7: Both calculate the total value of the expression, multiplying the coefficients to their factors and adding up the resulting plots. Whoever gets more value, wins the game.

Marino wanted to know from Iandu if some of those games was learned at the school, with the mathematics teacher. Iandu answered that one or other game was learned really with the teacher in the classroom, when that was in keeping with the topic of the program that was being developed. But the teacher didn't allow the students to use the time of the class for the pupils to play amongst themselves. He who wanted to play using the games had to do it out of the class.

6. A VISIT TO THE MEDICINE MAN

On Sunday in the morning, Marino and Lalo came out to fish, taken by the mayor and Iandu. In a few time they were throwing their nets in the water and picking fish, turtles, branches, snakes, slime and other elements that populate the underwater world.

At certain moment Marino entered the water, after seeing Iandu to do the same. But, then, he heard a scream of Mr. mayor: "Take care, Marino! Take care with the candiru!"

Candiru, according to the explanation that Marino started to hear from Iandu, is a small fish that penetrates the imprudent people's holes. It can penetrate even in the urethra. The worst thing is that, as people tell, once it enter there, the candiru opens the grill and gets hooked. There is nobody capable to do it to leave.

The fishery was worthwhile and was entertaining, but Marino dawned the following day with the body full of itches, what motivated Iandu to invite him to go to visit the medicine man house. This one was not as the medicine men of old times, who were devoted to manipulate supposedly supernatural elements. The medicine man now, although he practiced some inoffensive divinatory arts, served the population with the skill of prescribing natural medicines for the most diverse diseases that could appear in Guarana.

Arriving at the house, Marino was soon telling about his itch, which even he did not need to have done, because he scratched the whole time. The medicine man issued a laughter.

18

"Why are you laughing?", Marino asked him. "It is because this is very simple", the medicine man answered. "As soon as you return home, squeeze one or two cloves of garlic and rub in the areas of the body where you feel it was scratching. If you doesn't want to pass the entire day with the smell of garlic, then make this at night, when you are going to sleep. Tomorrow there won't be more itch".

After they to be half hour more in the medicine man house, talking about the traditions of the tribe, Marino and Iandu remunerated him with some coins and returned to the mayor's house.

At night Marino asked Iandu to teach him one more game, in what he was not assisted, because, according to Iandu, the mathematics teacher would teach a game in the Monday class and Marino should be present. Iandu already knew the game, but he didn't want to advance anything in that moment. He didn't want to destroy the flavor of the novelty in his teacher's class.

7. THE GAME OF FRACTIONS IN THE DOMINO

On Monday, the first class was mathematics and the teacher has done a small explanation before teaching to the class the "game of fractions in the domino". He explained that mathematics is a great sweet orange, composed of many buds, but that there are prejudiced people who rejoice in spreading around the world that it is a sour orange.

He told that the names of the slices of the mathematical orange are arithmetic, geometry, algebra, trigonometry, analysis, logic, probability, statistics, computation and information theory. He said also that the last person who got to know well all of the existent slices in his time was the French engineer and scientist Jules-Henri Poincaré (one pronounces it "pooancorhay"), who died in 1912. After that time the several topics of this science were enlarged in such a way that nobody more could dominate them entirely.

Soon afterwards the teacher started to expose to the class the steps of the announced game, not without before alerting for the school norm of work, that recommended one didn't use the time of the class to play with the number games, being this time able to be, at the most, used for the learning of those games, in way to there not to be interference in the course of the content. Some students who already knew the game were requested to render attention and to have some patience while the others heard the explanation. Several of them put their dominoes on the desk and the teacher began the exhibition.

Step 1: One puts a domino on the table. After shuffling the

bones, each of the two participants of the game withdraws two bones. They should stay with the face turned down in front of each player until the moment of turning them for the verification of the values.

Step 2: After turning the bones, one considers in each the half turned for the participant as being the denominator and the other half as being the numerator, each bone working as a fraction. If, in the sequence in that the bones of each participant appear, the first bone represents a fraction larger than the second one, the participant subtracts one fraction from the other, writing down the result. If, to the opposite, the first fraction is smaller than the second one, or equal, the participant adds the two fractions, writing down the result.

Step 3: The participant who obtains the larger result withdraws two more bones, repeating the same process adopted for the first two. If the result of that second picking goes larger than that of the first, the participant withdraws two more bones, repeating the process once again. If, to the opposite, the result goes smaller than the result of his previous picking, he should pass his turn to the other participant, who will proceed according to the rule described already for his opponent. Each participant's last result is being worth as his balance. When the last two bones are removed and the calculations regarding that step is made, the participant who on that moment has the larger balance wins the set. Any bone that appears with zero in the denominator is changed immediately for another. In there not being more bone to remove or to change, or just one remaining, the game closes up.

For making the game most agile one should take into

account the following observations:

A) The common denominator for the comparison, the sum and the subtraction in each step will be obtained in the following ways: (1) in the case of the couples of numbers 1_1, 1_2, 1_3, 1_4, 1_5, 1_6, 2_2, 2_4, 2_6, 3_3, 3_6, 4_4, 5_5 and 6_6 one takes, among the two numbers, what is multiple of the other (if one is larger than the other, the largest of the two is taken, if equal, the proper one is taken); (2) in the case of the couples 2_3, 2_5, 3_4, 3_5, 4_5 and 5_6, which are co-prime numbers (HCF is 1), one takes as multiple the product between the two; (3) in the case of the couple 4_6 one should have in mind that the lower common multiple (LCM) is neither the product nor the larger number, but, yes, the number 12, and this will be the denominator.

B) The comparison of the results between one and another set will demand calculations with more varied denominators and sometimes much larger than those mentioned above. If the calculation of the LCM takes the game to be very slow, the most suitable is to find the common denominators for simple product, unless it is noticed that a number is multiple of the other, when then this multiple is taken.

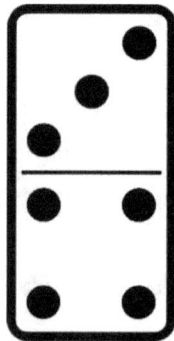

In this position,
one has $^3/_4$.

22

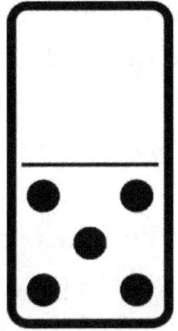

Here, we have $^0/_5$.

This stone has zero in the denominator. One excludes it of the match.

When returning to the house, Marino found Teacher Lalo in the desk, registering his impressions and discoveries on the life of Guarana. Marino summoned him immediately to buy a domino. "Eh! why do you want a domino?", Lalo asked. "Ah, dady, did you know that each bone of the domino is a fraction? A bone with a couple and a trio can be understood as two thirds", Marino answered. "Ah, so it is a new game that you have learned!"

The following day, in the morning, Lalo bought a domino and, soon, he and Marino were practicing the game of fractions. Again, Marino won the most of the sets. Lalo went out with the expected argument: Marino had more luck than him.

Lalo asked Iandu if some of the number games practiced in Guarana was done with deck of cards. The answer was negative and Iandu complemented explaining that the deck of cards game, for it to be considered bad luck game par excellence, it was discouraged in the city. Lalo said he have asked the question because he had some expectation that there did not be games with the deck of cards. Marino preferred to say that he have appreciated that "par excellence", of the Iandu answer.

8. THE GAME OF THE PROPORTIONS

Marino wanted to know from Iandu if there were no more games with dominoes. Iandu said that there was, yes, another game, but that it was more difficult than the one of fractions. That didn't worry Marino, who wanted to learn the new game soon.

Before beginning the steps of the game, Iandu made some explanations. First of all, he who is in the fifth grade or already has passed of it certainly knows the notion of fraction, but here it would be necessary to know the definitions of proper fractions and equivalent fractions. And he recalled that proper fractions are the ones that have numerator (number on the line) smaller than the denominator (number under the line). Equivalent fractions are the ones that have the same value, although written with different terms. For instance, 1/2 and 3/6 are equivalent, because both have value half of a whole.

After those notions, it is necessary to understand the meaning of the word reason, which is one of the manners of conceiving the division. A fraction is, therefore, a reason and the value 1/2 can be understood as the reason "1 for 2". The fraction 3/6 one understands it as the reason "3 for 6". The idea of proportion appears when we wrote the equality of two reasons, like 1/2=3/6. It is then read like "1 is to 2 as 3 is to 6" or, simply, in this case, "a half is the same as three sixth". The fundamental property of the proportions is that "the product of the means is the same as the product of the ends". This means that, in 1/2=3/6, we have 2x3=1x6.

25

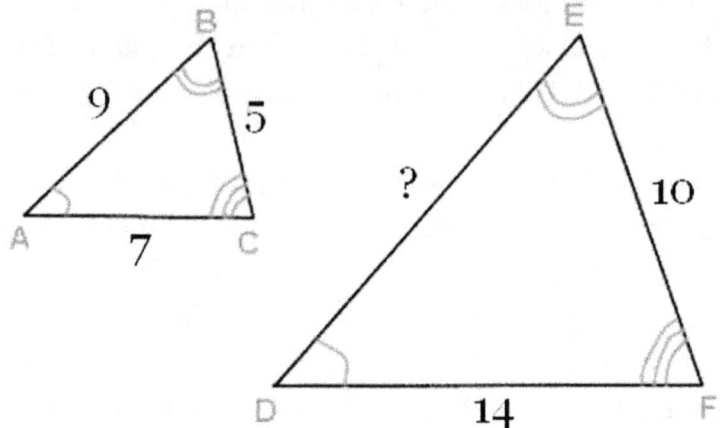

Figures with sides in proportion

After those explanations of Iandu, Marino could meet the game.

Step 1: The player A chooses, in ascending order, two different numbers between 0 and 12 (including these). Such numbers will form a fraction. He writes the numbers. Example: 4 and 11, forming the fraction 4/11.

Step 2: He chooses now 3 numbers between 1 and 6 (including these). He writes them. Example: 1, 3 and 6.

Step 3: He withdraws a bone of the domino now and writes its values as a fraction, with numerator inferior (or equal) to the denominator. For instance, the bone with couple and quintet will be 2/5. He equals this fraction to the fraction obtained in the step 1, finishing with interrogation. Example: 2/5=4/11?

Step 4: He applies the rule of proportions (property) to that equality. If it works (the product of the means is the same as the product of the ends), the player A wins 1 point and returns to the step 1, otherwise, the sequence of steps continues .

Step 5: The two reasons do not form a proportion. He chooses and writes, now, "numerator" or "denominator".

Step 6: He chooses, and writes, if he will "add" or "decrease", using the numbers of the step 2.

Step 7: He chooses one of the three numbers of the step 2 and makes the operation of the step 6 in the term chosen in the step 5, writing the new equality, that is to say, if the player chose 1 and he has written "denominator" and "decrease", he will have 11-1=10 and the new equality will be "2/5=4/10?" (the "11" did become "10").

Step 8: He applies the rule of proportions. If it works, the player A wins 1 point and writes it down, returning to the step 1. Otherwise, the round passes to the opponent, the player B. In the example, of 2/5=4/10, we have: 5x4=2x10. This means that the fundamental property is worth here. The player A wins 1 point.

The participant who first adds 4 uninterrupted points or, in another way, 7 points in interrupted playings, this one wins the set.

The game of the proportions is a long game, of many steps, and Marino did not believe that he could memorize it just for playing sometimes, that is why he asked for Iandu to write in a notebook paper the several stages of the game so that he could keep it in the luggage and take home in a registration safer than the mere capacity to remember. He also increased:

- I think you will need to write for me other games that I go learning. They are getting difficult.

- None of this - Iandu said, - there is other much simpler games that you didn't learn yet.

- But we won't trust so much - Marino returned. - The amount of games is big and the time of my stay here with my father is very small.

Iandu has taken Marino to practice that game of the proportions having as partners the mayor, the woman of this, that is, the mother of Iandu, and, later, Amana. Of this last one, Marino won all of the sets, so much that she decided to teach him a new game. It was the game of the decimals.

9. THE GAME OF THE DECIMALS

This game is suitable to the learning and the practice of the operations of addition and multiplication of numbers with decimal point. Here also the big die is used for the generation of the involved ciphers. Amana went indicating one by one the steps of the game.

Step 1: The player A takes in the big die three numbers, each with four ciphers. Then, for each of those numbers of four ciphers, the player B is going withdraw for A a number of one cipher, of which the remainder of the division by 4 is taken. It should be noticed that if the number is already smaller than 4, the rest of the division is the own number. For instance, if the number is 3, the division for 4 gives quotient zero and remainder 3, but if the obtained number is 9, for instance, the remainder of the division for 4 is 1, that is to say, from 9 the number 4 is subtracted twice and the remainder 1 is taken.

Step 2: The three remainders in the numbers withdrawn by the player B will be used to indicate the decimal order in that the decimal point should be put. If the first remainder is 3, the point put in the first number of 4 ciphers obtained for A should leave 3 ciphers to the right. If the number was 5213, it will be 5.213. If the second remainder is 1, the second number of 4 ciphers will be with one cipher after the point. Like this, if the number is 0297, it will be 029.7. If the third remainder is 2, two ciphers are left in the third and last number. The number 3741, for instance, will be 37.41.

Step 3: The player B takes in the big die three numbers of

29

four ciphers for himself and the steps 1 and 2 are repeated now relatively to him.

Step 4: The participants A and B add their numbers, following the norm of putting point under point in the frame of the account.

Step 5: The player A obtains with the big die a number of so many ciphers how many the ciphers after the point in the total obtained with the sum of his three ciphers are. This new number should begin with a no-null cipher. He multiplies it now by the total. From this product he subtracts the largest of the three numbers now initially added, taking into account the position of the point. (To decrease, he fills out with zeros the empty orders.)

Step 6: The player B repeats the procedure of A in the step 5 regarding his own total.

He who obtains the larger result wins the set.

A complete example is the following: The player A obtained the numbers 5213, 0297 and 3741 and, the remainders in the moves of B being equal to 3, 1 and 2, those numbers became 5.213; 029.7 and 37.41.

The values of B were 2318, 4977 and 5026. The ciphers that A took for forming the remainders of the division for 4 were 5, 8 and 2. The remainders were, obviously, 1, 0 and 2, so that for the three numbers the new formats 231.8; 4977 and 50.26 resulted. In the second number, the point is understood after the last cipher.

The total obtained for A was 72.323 and, for B, 5259.06.

Applying the following step, the player A takes in the big die the number 612 and multiplies it for 72.323. B takes the number 45, multiplying this value soon afterwards for 5159.06. The numbers obtained are 44061.676 for A and 236657.7 for B. Now, one makes 44061.676-37.410 = 44024.266, for A, and 236657.7-4977.0 = 231680.7, for B. Like this, B wins the set.

In the absence of a big die the game can be practiced with the use of a domino. In this case, the numbers to be considered are those that appear in the superior part of the bone, since it is put longitudinally in relation to the player who obtains it. Here, one should obviously to restore each bone before each obtainment. There is the disadvantage of one does not count on the ciphers 7, 8 and 9, but this should not be reason of impediment. Already a common die, of six faces, is not convenient because of the lack of zero.

10. THE GAME OF THE DIVISION

When Marino presented Lalo the game of the decimals, Lalo wanted to know if he had learned some game with the operation of division. Marino then noticed that no game had been introduced him with the use of that operation. He ran to know of Amana if there was such a game, but she sent him to Iandu. He is who knew well the game of the division.

Iandu caught the big die and began teaching Marino the steps of the new game.

Step 1: Each of the two participants raffles for his opponent four ciphers for dividend and two more for divisor. The first cipher should not be zero, other number being raffled if this appears. The two ciphers of the divisor should be different from zero. One of the participants (A) observes the other (B) to apply the process until the end. Then the roles are inverted.

A) 3158 |23   B) 2501 |39

Step 2: The participant A indicates the ciphers for the beginning of the division and researches the first cipher of the quotient, what should be made in the following way: He writes the multiple of 10 that more approaches the divisor (in the example it is 20) and goes multiplying whole values until the result is the largest number inferior or equal to the part indicated in the dividend for the operation.

32

```
31'58 |23      20     31'58 |23        A | B
        x 1    -23   1                1
        20     085
```

Step 3: The product found in the step 2 is written under the dividend and the operation continues. The player A wins 1 point, what is written down to the side. If, in any step, the result throw for A goes incorrect (a remainder larger than the divisor, or a remainder smaller than zero, i. e., subtrahend larger than the dividend), he will mark 1 point for B. If, to the opposite, he distrusts that the cipher of the last multiplication will produce incorrect result, he asks "authorization" to alter the value to be used as result. He marks 1 point then for B. If his alteration produces correct result, he marks 2 points for himself. If it produces incorrect result, he marks 1 more point for B.

```
20  20 ... 20      A | B      3158 |23
x 1  x 2   x 4     1   1      -23   13
20  40    80       2          085
                               -69
```

(The player A distrusted that the number 4 would give incorrect result and asked B for authorization to alter that value. As he got right, he marked 2 points, compensating the point that he marked for the opponent and registering his earnings. If he had made the operation with the value 4, verifying the mistake in the remainder, he would have marked 1 more point for B, and then he would repair the result of his operation, without marking any other point.)

Step 4: The continuation will be (supposing that A in this turn does not distrust the incorrect result):

```
3158 |23      20  20  ... 20        3158 |23
-23     138   x 5 x 6    x 8        -23     137
085           100 120    160        085
-69                                 -69
 168                  A | B          168
-184                  1   1         -161
                      2   1          007
```

Step 5: Finished the count, the participant A makes the proof of the operation and, if the final result is correct, he marks 1 point for himself. If it is wrong, he marks l point for B.

```
Proof: 137....... quotient   3151        A | B
       x 23....... divisor    +7         1   1
         411                 3158        2   1
         274_                            1
        3151
```

Step 6: The player B begins his count, and his multiple of 10, in the example given in the sequence, will be 40 (39 is closer to 40 than it is to 30; whenever the digit is equal or superior to 5, the rounding is upward).

```
250'1 |39     40  ...  40          A|B
-234     6    x 1     x 6          1   1
0161          40      240          2   1
                                   1   1
```

```
2501|39                              A | B
-234  64                             1   1
0161                                 2   1
-156                                 1   1
005                                      1


Proof:        64        2496        A | B
             x 39        + 5         1 1
              576        2601        2 1
              192                    1 1
             2496                    0 1
                       Balance:      4 4
```

He who possesses more points, in the final counting, this one wins the game. In the example, the match finished in tie.

11. THE GAME OF THE DECOMPOSITION

Iandu asked Marino if he wanted to learn the game of the decomposition.

- Is it the game of the decomposition in prime factors? - Marino asked Iandu.

- Quite so - Iandu answered.

- I am really needing to train that thing of prime numbers. Let's see how it is.

The steps of the game began to be shown. Before this, Iandu reminded that it is necessary to know the process of decomposition in prime factors and to know that the first prime numbers are 2, 3, 5, 7, 11, 13, 17, 19, etc. Prime number, he reminded, is the natural number that has exactly two distinct divisors. He drew the attention also for the fact that this game, as several others practiced in Guarana, does not request the fixed number of two players, being possible to play it with three, four or more people.

Step 1: Each participant draws in the big die a number of two ciphers, the first of them being different from zero. If the number obtained is prime, one launches the big die once again and extracts a cipher, which should be added to that prime number. If the result still is prime, one takes with the big die a new cipher, which now should be subtracted. Continuing prime, one extract other cipher and adds it, alternating like this the operation while it is necessary.

Step 2: Each participant makes the decomposition in prime factors of the number that resulted to him.

36

Step 3: To each distinct prime number in the decomposition 1 point is attributed. The prime numbers that are repeated one, two or more times are to value 2 points each.

Step 4: The result obtained with the counting of points of the step 3 is multiplied by the cipher that is in the position of the dozens of the number that was decomposed.

He who obtains larger result with the operation of the step 4 wins the departure. If there is a tie here, he who has more different prime numbers will win.

Example: The participant A obtains the number 27 and the participant B obtains 41. As the number of B is prime, he draws a cipher and sums. He took the cipher 1, what will leave him with 42. The factorization of those numbers will give 27=3*3*3 and 42=2*3*7. The points of A are 2+2+2 and the ones of B will be 1+1+1, according to the step 3. Now A multiplies his total of points, 6, for the number 2, cipher of the dozens in 27, obtaining the total 12. The player B multiplies his total, 3, for the number 4, dozen of 42, obtaining result 12. As a tie happened, one looks as the decomposition. B has more distinct numbers, therefore, B wins the match.

12. THE GAME OF THE MULTIPLICATION OF FRACTIONS

On Thursday in the night Iandu informed Marino that he would teach him the last game, that dealt with the multiplication of fractions and was done with the domino. There was still another game that Marino did not know, but that one of the mathematics teacher had said that he would teach it in the Friday class. The steps of the game of the multiplication of fractions were the following ones:

Step 1: Each player draws two bones of the domino and leaves them in longitudinal position. Each player turns his bones and multiplies the fractions represented through the values enrolled in them. If some bone appears with zero (white) in the denominator, the player draws other and substitutes the old, before multiplying. That bone with the zero is excluded of the match.

Step 2: The player who is with the larger result draws another bone and multiplies it for that result. In case of tie, one removes the new bones simultaneously.

Step 3: One turns to the beginning of the step 2. When the bones reach the end the match finishes and the winner is he who has the larger result. In all of the steps, one should simplify the fractions in the own product.

13. THE GAME OF THE SUM OF WHOLE NUMBERS

On Friday Marino and Iandu entered the class taking each his big die. Mathematics was the third class, which has arrived very quickly, and the teacher began the work making a lecture, as it was his habit. He reminded initially that Teacher Edmond Smith when was in Guarana left the practice of the number games among the people of what was then a village, but they were some few games, some three or four. Once the practice became a tradition, the Guaranaensis were creating other games, arriving to the order of dozens of them.

The teacher was getting more exalted when started to discourse on the value of the knowledge. "The knowledge acquired for the humanity should not be exclusive property of Europeans or Asiatics. The Indians from America need to have access the the whole knowledge of the world. It is a weapon that they cannot refuse to us." To illustrate that the investment in the human being is worthwhile, he said that several Guaranaensis exercise professions of the exact sciences in other cities and even in other States, there being even academical teachers of Mathematics and Physics researching in public institutions.

Soon afterwards the teacher started to explain the steps of the game that he had promised to teach on that day.

Step 1: The big die is taken and the drawing of the values begins, one observing their color. The player A launches the big die and obtains, for instance, the number 3, in red. He writes that number as -3 in a column topped by the word "points". To the right of the column a second column is topped by the word "balance". Under this word he also writes that number.

Step 2: The player B repeats the path of the player A, drawing a number and writing it under the column that will indicate his points. If the number obtained for B is larger than that of A, this player subtracts, when his time arrives. If the number of B is smaller, or equal, the player A will add. If the number in this case is +1 (the "1" appeared with the color blue, or black) the player B will mark +1 under the column "points" and +1 under his column "balance". As +1 is larger than -3, it will be the turn of the player A to subtract his next number to his balance.

Step 3: The participant A draws his number. Supposing that the value is -4, he will subtract it of the balance -3, obtaining the result +1, because -3-(-4)=-3+4=+1. This will be written as "-(-4)" under the column "points" and "+1" under the column "balance".

Step 4: The player B now should add in his balance, because that "-4" is smaller than his last number obtained, which was +1. If he obtains the value -5 now, then the disposition in the columns will be "+(-5)" under the column "points" and "-4" under the column "balance", because +1+(-5)=-4.

The game is continued until that each participant has taken seven values of the big die. Who has the larger balance wins the match.

Possible disposition of the game until the end of the match:

A				B	
points	balance			points	balance
-3	-3			+1	+1
-(-4)	+1			+(-5)	-4
+(+2)	+3			-(-2)	-2
+(-1)	+2			-(+4)	-6
-(+6)	-4			-(+1)	-7
+(-6)	-10			+(+5)	-2
-(-9)	-1			+(+2)	0

In the match above the player B wins, because zero is larger than -1.

Remarks: A) the comparison to know if one should add or subtract is always on the numbers drawn of the last play of each and not on the balances. B) The fact of one to add or to subtract does not represent advantage or defeat, since one is dealing with addition of relative numbers. What does count for the victory is the final balance.

14. LEAVING OF GUARANA

Returning of the school, Marino saw in a window a kind of parrot that he never had saw before. He asked Iandu on what type of bird it was.

Now, it is a purple-maitaca, Iandu said.

- Maitaca? But is not maritaca the name? - Marino wanted to know.

- Just a minute, if you knew, why did you ask? - Amana intervened.

- He found strange the form of the name, Amana. He didn't know a purple-maitaca, but he knew the word maritaca. Is not this, Marino? - Iandu wanted to confirm.

- Yes, it is this. - Marino answered. - Is maitaca then the same thing that maritaca?

- It is, yes - Iandu said. – Your people there pronounces with one "r" in the middle and we here in the Brazilian Amazon pronounce without that "r". By the way, this is not a type of domestic parrot. I don't know what that one does in that window.

Marino and Teacher Lalo would leave the following day, Saturday, in the morning. They were felt for they could not be for seeing the Kuarup, the return of the sun party, indigenous traditional ceremony that would be accomplished the following week. But the objective of the trip to Guarana was done. Marino learned and wrote down in his notebook nothing less than ten number games and with this there would be a lot of amusement with the group of his school. Besides he had learned how to build the big die, an innovation for his friends. Lalo, for his time, with that material registered, had many things to write in his academic work.

Iandu told about a dream that he had on that last night: "I dreamed that when the day of leaving arrived all of you decided to stay, but in a little while you were flying, going away."

Saturday, in the morning, which would be the day of the departure, the weather was worse and Marino was wakened up by the noise of the rain. In the hour of the breakfast, Lalo said that the rain could delay the trip and everything that they had to do was to wait and to try to take advantage of the spare time of that unexpected extension.

Marino that morning passed playing with the mayor, Mr. Tieh. First they played a lot of matches of the game of the multiplication of fractions, then, of the sum of whole. Curious thing: so much in one as in the other game, the mayor won the great majority of the matches, having even won the last black one.

It was the time of Lalo to mock the son:

- Now the mayor is getting the revenge of a lot of matches that you won of me.

- It is, one arrives in a day in that the hunter becomes hunt, - Marino sentenced, discouraged.

After lunch, the clouds left and in a little time the earth was dried. In the afternoon the airplane arrived, with several hours of delay, very justified, and then the teacher and his son left the city of Guarana, answering to the nodding of the mayor, of the woman of this, of Amana and Iandu and of several girls of the school who went to the airport to dismiss of Marino.

Marino entered a space rocket, that dove in a cylindrical and silver tunnel, and in a little time he was on the surface of the Moon. There, Marino found a soccer ball whose buds contained drawn illustrations of boys and girls. He kicked it lightly. With the rarefied air and the little gravity it arose much more than what was expected. But before it won height, the boys and girls drawn in it won life and went jumping from the ball to the soil. Then they formed somewhat a platoon, screaming rhythmically: "He doesn't win just of the mayor! He doesn't win just of the mayor!" Then, more ahead, the children of the platoon went becoming rufescent tiger-herons and seagulls, which took off and steered towards the Mare Tranquillitatis, that now was a sea of waters and not only of sand. The last boy to transform in cracid was turned back and said: "This sea was already a pororoca, a tidal bore."

In that moment Marino decided to look at the sides and then he noticed that there was a formation of dunes and that he was not at the surface of the Moon. It was the city of Cabo Frio. There in front it was the fort. Therefore the Mare Tanquillitatis was not dry, Marino thought. But, again, a group of boys, the same group who he saw in the Moon, passed running and screaming: "He doesn't win of the mayor!" Marino threatened to run behind the boys, trying to swear at them. On that instant Lalo held him: "What happened, son?" Marino was afraid: "Just a minute, daddy, I am not in Cabo Frio? Just now I was in the Moon!" Lalo mocked: "It seems that the jolt of the airplane had drastic effects on you. By the way, you should be affected. You won of all, but you didn't win of the mayor." Marino answered angry: "Even you, daddy?"